YOUR KNOWLEDGE HAS VALUE

- We will publish your bachelor's and master's thesis, essays and papers

- Your own eBook and book - sold worldwide in all relevant shops

- Earn money with each sale

Upload your text at www.GRIN.com and publish for free

Bibliographic information published by the German National Library:

The German National Library lists this publication in the National Bibliography; detailed bibliographic data are available on the Internet at http://dnb.dnb.de .

Imprint:

Copyright © 2016 GRIN Verlag, Open Publishing GmbH
Print and binding: Books on Demand GmbH, Norderstedt Germany
ISBN: 9783668476783

This book at GRIN:

http://www.grin.com/en/e-book/366729/antimicrobial-and-phytochemical-analysis-of-lime-juice-and-different-types

Prem Jose Vazhacharickal, Sajeshkumar N. K., Jiby John Mathew, Steffy
Maria Augustine

Antimicrobial and phytochemical analysis of lime juice and different types of honey. An overview

GRIN Publishing

Antimicrobial and phytochemical analysis of lime juice and different types of honey: an overview

Prem Jose Vazhacharickal, Sajeshkumar N.K, Jiby John Mathew and Steffy Maria Augustine

ACKNOWLEDGEMENTS

Firstly we thank **God Almighty** whose blessing were always with us and helped us to complete this project work successfully.

We wish to thank our beloved Manager **Rev. Fr. Dr. George Njarakunnel,** Respected Principal **Dr. Joseph V.J,** Vice Principal **Fr. Joseph Allencheril**, Bursar **Shaji Augustine** and the Management for providing all the necessary facilities in carrying out the study. We express our sincere thanks to **Mr. Binoy A Mulanthra** (lab in charge, Department of Biotechnology) for the support. This research work will not be possible with the co-operation of many farmers.

We are gratefully indebted to our teachers, parents, siblings and friends who were there always for helping us in this project.

Prem Jose Vazhacharickal*, Sajeshkumar N.K, Jiby John Mathew and Steffy Maria Augustine

Table of contents

Table of tables

List of abbreviations

%	: Percentage
°C	: Degree celeseus
µg/µl	: Micro gram per micron liter
AIDs	: Acquired immune defficency syndrome
Aw	: Water activity
bp	: Base pairs
$C_2H_3CuO_2$	: Copper acetate
Ca	: Calcium
CDC	: Center for disease control and protection
cm	: Centimeter
cm	: Centimeter
DNA	: Deoxy ribonucleic acid
E.coli	: *Escherichia coli*
Fe	: Iron
$FeCl_3$	: Ferric chloride
GC	: Guanine and cytosine
H_2O_2	: Hydrogen peroxide
HCl	: Hydrochloric acid
HDL	: High density lipoproteins
HIV	: Human immunodefficency virus
HNO_3	: Nitric acid
K	: Potassium
KDa	: Kilo Dalton
Kg	: Kilogram
L	: Liter
LDL	: Low density lipoprotein
M	: Meter
Mg	: Magnesium
$MgCl_2$	: Magnesium choloride
MGO	: Methylglyoxal
MHA	: Muller hinton agar
MIC	: Minimum inhibitory concentration
Na	: Sodium
NaCl	: Sodium chloride
nm	: Nano meter

P	: Phosphorous
RNA	: Ribonucleic acid
SE	: Standard error
TSS	: Total soluble solids
UTI	: Urinary tract infection
WBC	: White blood cells
µg	: Micro gram
µl	: micro liter

Antimicrobial and phytochemical analysis of lime juice and different types of honey: an overview

Prem Jose Vazhacharickal[1]*, Sajeshkumar N.K[1], Jiby John Mathew[1] and Steffy Maria Augustine[1]
[1]Department of Biotechnology, Mar Augusthinose College, Ramapuram, Kerala, India-686576

Abstract

Citrus, one of the major genes of Rutaceae family and most economically important fruit tree and widely cultivated throughout the country. The Citrus have high nutritionl value and medicinal value. Honey and lemon-honey are traditional remedies in the Middle East and China and for many centuries and have been used in the treatment and prevention of the common cold and various upper respiratory tract infections. Three types of honey were collected; 'Cheruthen'-produced by bees belongs to the *Trigona irridipennis* species; Vanthen'- produced by bees belongs to the *Apis indica* species; 'Kattutthen'- produced by bees belongs to the *Apis dorsata* species. The antibacterial activites of honey samples and lime juice were tested aganist Bacillus, Klebsiella, *E.coli*, Staphylococcus and Micrococcus. The result showed that the samples have differnt antimicrobial activity. Antimicrobial activity of Cheruthen against Klebsiella species showed a zone of inhibition of 10.1 ± 0.73 mm, when 100 µl of Cheruthen is applied. When 200 µl of Cheruthen is applied the zone of inhibition was 30.1 ± 0.23 mm. Antimicrobial activity of Cheruthen against E.coli showed a zone of inhibition of 10.1 ± 0.13 mm, when 100 µl of cheruthen is applied. When 200 µl of cheruthen is applied the zone of inhibition was 30.2 ± 0.23 mm. Also the phytochemical examination of lime juice and honey samples showed that different types of phytochemical substances are present in both lime juice and different types of honey samples. Further studies are required to reveal the role of each phytochemical and its contribution to the antimicrobial properties of the samples included in this study.

Keywords: Anti-oxidant; lime juice; Phytochemical analysis; Honey samples.

1. Introduction

The universal role of plants in the treatment of disease is exemplified by theiremployment in all the major systems of medicine. Natural products, either as pure compounds or as standardized plant extracts, provide unlimited opportunities for new drugs because of the ready availability of chemical diversity (Cos et al., 2006). Honey and lemon-honey are traditional remedies in the Middle East and China and for many centuries and have been used in the treatment and prevention of the common cold and various upper respiratory tract infections (Molan, 1992; Zulma and Lulat, 1989).

Honey is defined as the sweet substance produced by honey bees from the nectar of blossoms or from secretions of living plants, which the bees collect, transforms and store in honey combs. Honey bee produces dense stable energy food called as nectar which ripened into honey.It also has antioxidant and antimicrobial properties. It is a complex mixture of inverted sugar concentratedsolution that also contains some other carbohydrates, aromatic substances, waxes, minerals, pollen grains, pigments, organic and aminoacids (Arshad, 2003). Honey contains approximately 80% sugar by weight, which is comprised of four main sugar molecules (fructose, glucose, maltose and sucrose), and with many others in lower quantities. The acids present in honey also help to restrict microbial growth. Complex mixtures of acids, particularly gluconic acid, contribute to low acidity and low pH between 3.4 and 6.1 (White, 1979). These characteristics alone make all honeys unsuitable tosupport the growth of microorganisms and explain why honeys destined for human consumption rarely spoil during storage in the home.The use of honey iswide spread among consumers all over the world and used in baking, cooking as a spread on bread and also as an additive in various commercial beverages.

Honey is a supersaturated solution of sugars, with low water content, and the binding of water molecules to sugars makes them unavailable for microorganisms. The availability of free water is expressed as water activity (Aw); pure water is Aw 1.00 and blood 0.99. Most honeys possess an Aw of approximately 0.6, and many microbial species require Aw between 0.94 and 0.99 to grow (Molan 1992; Chirifeet al., 1982; Gisleneet al., 2000). For a long period of time, plants have been valuable source ofnatural products for maintaining human health. The use of plant extracts and phytochemicals, both with known antimicrobial properties, can be of great significance in therapeutic treatments (Seenivasan et al., 2006). Many plants have been used because of their antimicrobial traits, which are due to compounds synthesized in the secondary metabolism of the plant (Al-Shuneigat et al., 2005). Hence the essential oils and other extracts of plants have evokedinterest as sources of natural products. They have been screened for their potential uses as alternative remedies (Tepe et al., 2004; Dorman and Deans, 2000).

The genus Citrus, which includes mandarin, orange, lemon, grapefruit and lime, has high economic and nutritional value. This genus belongs to the subfamily Aurantioideae, which is one of the 7 subfamilies of the family Rutaceae. Therefore, phylogenetic study of both the genus Citrus and of the subfamily Aurantioideae is important (Penjor et.al 2013). The Aurantioideae consists of 2 tribes with 33 genera (Swingle et al., 1967). These 2 tribes are each composed of 3 subtribes: the tribe Clauseneae, which includes Micromelinae, Clauseninae, and Merrillinae; and the tribe Citreae, which includes Triphasiinae, Citrinae, and Balsamocitrinae. None of the Clauseneae species develop axillary spines, and the odd-pinnate leaves have alternately attached leaflets. The fruits are usually small and carry semi-dry or juicy berries, except in Merrilli (Swingle et al., 1967). In contrast, nearly all the species develop axillary spines in the Citreae. The simple leaves are easily distinguished from those of the tribe Clauseneae (Swingle et al., 1967). The genus Citrus belongs to the "true citrus fruit trees." The characteristics of Citrus species include asexual reproduction, high mutation frequency, and cross compatibility between species. Because of these characteristics, there is great morphological and ecological diversity among Citrus species Citrus species can be classified into 3 clusters: a citron cluster, a pummelo cluster, and a mandarin cluster.

Citrus is one of the most important commercial fruit crops grown in all continents of the world (Kanazeetal, 2008). Importance of citrus is attributed to its diversified use and growing world demand with about 102.64 million tones total world production and probably stands first largest among the produced fruit (Uchechi et al., 2010). The Citrus juices of lemonand bitter orange (*Citrus lemon* and *Citrus aurantium*) showed good antibacterial activities against gram positive and gram negative microorganisms (Manthey and Grohmann, 2001) and also obtained by Cvetnic 2004 who found that no inhibiting effect of *Citrus paradision* the growth of gram negative microorganisms. Lemon is an important medicinal plant of the family Rutaceae. It is cultivated mainly for its alkaloids, which are having anticancer activities and the antibacterial potential in crude extracts of different parts (leaves, stem, root and flower) of lemon against clinically significant bacterial strains has beenreported (Kawaii et al., 2000). Lime is a small shrub like tree ranging from 3.5 to 9 m in height and 2.5 to 7.5 m in width. The fruit is typically round, green to yellow in colour and about 3-6 cm in diameter.

Citrus flavonoids have a large spectrum of biological activity including antibacterial, antifungal, antidiabetic, anticancer and antiviral activities (Burt, 2004; Ortuno et al., 2006). Flavonoids can function as direct antioxidants and free radical scavengers, and have the capacity to modulate enzymatic activities and inhibit cell proliferation (Duthie and Crosier, 2000). In plants, they appear to play a defensive role against invading pathogens, including bacteria, fungi and viruses (Sohn et al., 2004). Preparation from peel flowers and leaves of bitter orange (*Citrus aurantium*. L) are popularly used in order to minimize central nervous system disorders (Pultrini et al., 2006; Shahnah et al., 2007). In addition the fibre of citrus fruit also contains bioactive compounds, such as polyphenols, the most important being vitamin C (or ascorbic acid), and they certainly prevent and cure vitamin C deficiency the cause of scurvy (Aronson, 2001). The studies showed that essential oils, protopine, lactones, polyacetylene, hypericin and pseudohyperic compounds are effective toward various bacteria. Nevertheless, other active trepenes, as well as alcohols, aldehydes, and esters, can contribute to the overall antimicrobial effects of the essential oils (Keleset al., 2001).

The citrus peels are rich in nutrients and contain many phytochemicals they can be efficiently used as drugs or as food supplements too. Since there is an increase in the number of antibiotic resistance pathogens, food processors, food safety researchers, and regulatory agencies have been increasingly concerned with the growing number of food-borne illness outbreaks caused by some pathogens (Wilson and Droby 2000; Friedman et al., 2002; Soković et al., 2007). In general, flavonoids may contribute to fruit and juice quality in many ways, influencing the appearance, the taste and the nutritional value of the product from the plant (Hayes and Markova 2002).

1.1 Objectives

The objective of this study was

- A comparative study on the antimicrobial activity of various honey samples and lime juice.

- A comparative study of the chemical constituents like of phenol, flavanoids, alkaloids, glycosides, tannins, saponins and carbohydrates of various honey samples and lime juice.

1.2 Scope of the study

The scope of the study is to explore the antimicrobial activity of various honey samples and lime juice which could be used in many pharmaceutical formulations.

1.3 Taxonomical classification

Kingdom: Plantae-- planta, plantes, plants, vegetal Subkingdom: Tracheobionta -- vascular plants Division: Magnoliophyta -- angiosperms, flowering plants, phanerogames Class: Magnoliopsida -- dicots, dicotyledones, dicotyledons Subclass: Rosidae Order: Sapindales Family: Rutaceae - Rue family Genus: Citrus - citrus Species: *Citrus aurantifolia/Citrus maxima/Citrus limon*

2. Review of literature

Honey possesses antibacterial activityand is classified as peroxide and nonperoxide components. Data of origin and floral sources of honey antimicrobial activity against different bacteria species are summarized (Ahmad et al., 2006). It was concluded that inhibition of growth of bacteria is principally due to the peroxide effect, which is very common in honey wide world because it is a derivative compound from bee. Honey is a nutritive food used widely in the food industry, which provides energy to the organism due its high percentage of carbohydrates, which are easily assimilated (Ransom, 1937; Allen et al., 1991).

Honey is a natural product having carbohydrates and other compounds. Carbohydrates are mainly in the form of fructose and glucose while also present in the form of maltose, sucrose and other complex carbohydrates are in traces. Some vitamins are also reported in Honey samples and several other compounds in honey are phenolic contents, vitamins, catalase, pinocembrin, vitamin C and catalase. Trace amount of minerals was also found in honey (Bertoncelj et al., 2007). Honey possessesmany nutritional and photochemical importances due to the presence of these compounds. Which shows antifungal, antibacterial and antioxidant activities. The quality of honey depends on the flowers relatively long so that causes toxin infections in variability and availability. There are different diseases caused by bacterial and protozoan, European Foulbrood, American Foulbrood, which infects the honey bees, for curing these diseases the bee keepers mostly used different antibiotics (Abdul et al., 2012).

Honey is composed of about 183 components and is basically a solution supersaturated in sugars, of which fructose (38%) and glucose (31%) are the most important (Gheldof et al., 2002) the moisture content is about 17.7%, total acidity 0.08% and ash content is 0.18% (Nagai et al., 2006). In addition, there is a great variety of minor components, including phenolic acids, flavonids, the enzymes, glucoseoxidise and catalase, ascorbic acid, carotinoids, organic acid, amino acid and protein (Ferreres et al., 1993). However, the composition of honey varies depending on many factors such as the floral source, climate, environmental conditions and the processing it undergoes as pasteurization or storage (Gheldof et al., 2002; Azeredoet al., 2003). Honey is a unique food product containing bio active compounds derived from bees and plants. These bioactive compounds could be linked to the antimicrobial activity which has the capacity of destroying or inhibiting the growth of some pathogenic vegetative microorganisms (Allen et al., 2002; Nzeako and Hamdi, 2000; Chick et al; 2001).

Honey is gaining acceptance by the medical profession for use as an antibacterial agent for the treatment of ulcers and bed sores, and other surface infections resulting from burns and wounds (Mandal and Mandal, 2011). There has been much emphasis on moisture content, ash content, total soluble solids (TSS), acidity, pH, reducing, non-reducing and total sugars, fructose, glucose, diastatic activity, energy, minerals, microbial characteristics and organoleptic evaluation in the analysis of honey. Moisture is one of the most important characteristics of honey, having profound influence on its keeping quality and granulation of honey. The moisture content of honey is the range of 22.8 to 25.0% (Giuseppe et al., 2007). The determination of pH is important in honey in relation to darkening. As the pH increases the darkening of honey also increases, pH value of 4.10 in fresh honey, the protein content of honey as 0.169% (Bosi et al., 1978). Enzyme diastase (amylases) breaks down starch into simple sugars. It is very sensitive to heat and hence its activity is an indicator of improper heating. Therefore proper heating and storage is of almost importance to retain the market value of honey (Kandil et al., 1987). The values of calcium (0.5 mg/100g), phosphorous (16 mg/100g) and iron (0.696 mg/100g) present in honey.

Phosphorous (P), magnesium (Mg), iron (Fe), calcium (Ca), potassium (K) and sodium (Na) contents present in honey and reported that honey contained 20, 5.60, 0.76, 7.50, 75.40 and 7.10 mg/100g of these minerals respectively (Seymour, 1951). The antibacterial nature of honey is dependent on various factors working either phenolic compounds, wound pH, pH of honey and osmotic pressure exerted by the honey. Hydrogen peroxide is the major contributor to the antimicrobial activity of honey, and the different concentrations of this compound in different honeys result in their varying antimicrobial effects (Wahdan, 1998). The different floral sources may play important role in the antimicrobial activity of honey. Several authors reported that different honeys vary substantially in the potency of their antibacterial activity, which varies with the plant source. Thus, it has been shown that the antimicrobial activity of honey may range from concentrations < 3 % to 50 % and higher. The bactericidal effect of honey is reported to be dependent on concentration of honey used and the nature of the bacteria. The concentration of honey has an impact on antibacterial activity; the higher the concentration of honey the greater its usefulness as an antibacterial agent (Adeleke et al., 2006).

Antimicrobial activities of 10–100% (wt/vol) concentrations of new honey, stored honey, heated honey, ultraviolet exposed honey, and heated stored honey were tested against common human pathogens, including *Escherichia coli*, *Entrobacter cloacae*, *Pseudomonas aeruginosa*, *Shigella dysenteriae*, Klebsiella species, *Haemophilias influenza*, Proteusspecies, *Staphylococcus aureus*, *Streptococcus hemolyticus* group B, and *Candida albicans* (Al-Waili et al., 2005). Antimicrobial activity of honey was tested in acidic, neutral, or alkaline media. These were compared with similar concentrations of glucose in nutrient broth. The most sensitive microbes were *E. coli*, *Pseudomonas aeruginosa*, and *Haemophilias influenza*. Glucose showed less antimicrobial activity than honey, and many microbes showed positive culture even in 100% glucose (Lusbyet al., 2005). Heating to 80°C for 1 hour decreased antimicrobial activity of both new and stored honey. Storage of honey for 5 years decreased its antimicrobial activity, while ultraviolet light exposure increased its activity against some of the microorganisms. Antimicrobial activity of honey was stronger in acidic media than in neutral or alkaline media (Molan, 2001). Single doses of honey used to prepare the 60% concentration in nutrient broth were bactericidal for Pseudomonas aeruginosa and bacteriostatic for *Staphylococcus aureus* and *Klebsiella species* during certain periods. Local application of raw honey on infected wounds reduced redness, swelling, time for complete resolution of lesion, and time for eradication of bacterial infection due to *Staphylococcus aureus* or Klebsiella species. Its potency was comparable to that of local antibiotics. Honey application into infective conjunctivitis reduced redness, swelling, pus discharge, and time for eradication of bacterial infections due to all the isolates tested (Weston, 2000).

The beneficial role of honey is attributed to its antibacterial property with regards to its high osmolarity, acidity (low pH) and content of hydrogen peroxide (H_2O_2) and non-peroxide components that is the presence of phytochemical components like methylglyoxal (MGO) (Natarajan et al., 2001). The support for using honey as a treatment regimen for peptic ulcers and gastritis comes from traditional folklore as well as from reports in modern times. Honey may promote the repair of damaged intestinal mucosa, stimulate the growth of new tissues and work as an anti-inflammatory agent. Raw honey contains copious amounts of compounds such as flavonoids and other polyphenols which may function as antioxidants. Clinical observations have been reported of reduced symptoms of inflammation when honey is applied to wounds. The removal of exudates in wounds dressed with honey is of help in managing inflamed wounds (Blassa et al., 2006).

Citrus is a genus of family Rutaceae that comprise some 158 genera and 1900 species (Mabberley, 2008). It is mainly tropical to semi tropical in origin and is assumed to have originated from the region within Northeast India, South China, Indonesia and Peninsular Malaysia (Swingle et al., 1967). Citrus grows particularly well in areas where there is enough rainfall or irrigation to maintain growth and freezing conditions are not severe enough to kill the tree (Whiteside et al., 1998). Citrus is also one of the most important fruit crops in the world and its international production has reached 122 million tons (FAO, 2008).

Citrus fruits and juices are a great source of bioactive compounds including antioxidants such as ascorbic acid, flavonoids, phenolic compounds and pectins that are important to human nutrition (Ebrahimzadeh et al., 2004; Fernandez-Lopez et al., 2005; Jayaprakasha and Patil, 2007). The peel which represents almost one half of the fruit mass has the highest concentrations of flavonoids in the citrus fruit (Manthley and Grohmann, 1996, 2001; Anagnostopoulou et al., 2006). A wide range of DNA markers is available and has been used to study the classification of Citrus genus, and phylogenetic relationships within citrus and with related genera. These molecular studies have provided some insight to citrus phylogeny (Wali et al., 2013). The phylogeny and taxonomy of citrus fruit are complex, confusing and controversial due to the genetic heterogeneity of the genes, as wellas its polyembryonic nature and the long generation time needed to carry out selection and recombination (Swingle et al., 1967, Wicolosi et al, 2000).

Plants are the cheapest and safer alternative sources of antimicrobials (Adriana et al., 2007) .Citrus aurantifolia is a genus of flowering plants in the family Rutaceae (orange family) and a Common name for edible fruits of this genus and sometimes related genera. Lime (*Citrus aurantifolia*) juice has been shown to have both medicinal and cosmetic values (Kawaii et al., 2000). Fruit is reported to have antioxidant and anticancer properties. It has been used to tone and purify liver, in treatment of ulcers, arthritis, gout, and acne. Studies have shown that lime juice destroys both human immunodeficiency virus (HIV) and sperm cells. The high acidity of the lime juice is probably responsible for the destruction of the HIV and sperm cells. Its leaves and stems are reported to have antibacterial action. All parts of the plants of citrus species contain coumarins and psoralins (Moro et al., 2000).

Lemon juice has even been shown to be useful as an anti-HIV agent when applied vaginally in sexually active women. Another study reported significant larvicidal activity by a fresh lemon peel extract. Water containing lemon, however, was found to actually enhance the growth of *Pseudomonas aeruginosa* in one study (Ibrahim and Ogunmodede, 1991). Moreover, some lemon exporters spray the fruit with antimicrobial chemicals in order to kill *Vibrio cholerae*, *Penicillium digitatum*, *Botrytis cinerea*, and other microbes that may be contaminating the rind; this procedure indicates a lack of faith in the antimicrobial properties of lemon (Siddiqi, 1998; Brown, 1995; Michaud, 1999).

Lime juice, lemon juice, vinegar or acidic soft drinks in the belief that it may prevent pregnancy and or sexually transmitted diseases (Imade et al., 2005). Potentially, to be effective against HIV in vivo, women would need to apply a volume of neat lime juice equal to that of an ejaculate, and maintain this ratio vaginally for 5 to 30 minutes after ejaculation. However, data have suggested that this would have significant adverse effects on the genital mucosa, raising serious questions about the plausibility and safety of such a preventive approach. Lemon and lime juices have been reported to exhibit antimicrobial activity against Vibreo strains. The in vitroeffects of concentrated lime juice extract suggest that the juice may have anti proliferative effects on tumourcell lines (Gharagozloo et al., 2002). Citrus fruits are acidic fruits which contain healthy nutritional content that works wonders for the body. It acts as a number of mono and disaccharides (Giuseppe et al., 2007).

Citrus peel extracts showed a significant antibacterial activity against all the test organisms. *Citrus lemon* contains a very good antimicrobial activity. This antibacterial activity may be indicative of the presence of metabolic toxins or broad spectrum antibiotic compounds. The medicinal value of these plants lies in bioactive phytochemical constituents that produce definite physiological action on the human body (Akinmoladun et al., 2007). The preliminary phytochemcial investigation revealed the presence of various constituents of citrus peels. Different solvent showed different class of phytochemicals they showed the presence of flavanods and saponins. Antraquiones were completely absent in both the citrus peels. The presence of phenol further indicated that Citrus lemonand Citrus sinensis peels could act as antiinflammatory, anticlotting, antioxidant, immune enhancers and hormone modulators. *Citrus lemon* and *Citrus sinensis* peels have high quantity of saponin which has haemolytic activity and cholesterol binding properties. Therefore, in addition to their use as drugs, citrus peels can be used as a food preservative or even as foodsupplement as many literatures say that they are highly nutritive (Shale et al., 1999). The antimicrobial activities of citrus plants oil and extracts were investigated. Studied the effect of essential oils from *Citrus aurantium*, *Citrus limon*, *Citrus paradisi* and many other plant oils and extracts and found that the minimum inhibitory concentrations (MIC) were between 5-2% (v/v) and analyzed the action of combined spices including lemonin its mixture and found that the spices mixture were able to exert static effect on all assayed bacteria and also investigated the antimicrobial properties of lemonand found that lemonpossesses significant antimicrobial activity against *Staphylococus aureus*, Klebsiella, *Escherichia coli*, *Pseudomonas aeruginosa* and *Candida albicans* (Hammers et al., 1999).

The screening of antimicrobial activities of citrus extract on the tested bacteria used is usually expressed in millimetres (mm). The antioxidant activities of citrus flavonoids exhibited a potent antibacterial activity which is probably due to their ability to complex with bacterial cell walls and disrupt microbial membrane (Hayes et al., 2002). Differences in susceptibility between *Staphylococcus aureus*, *Pseudomonas aeruginosa* and *Proteus vulgaris* can be explained by differences in the nature and extent of cell membrane damaged (Ortunõ et al., 1999).

Many natural substances may play a fundamental role in the hostplant pathogen relationship the essential oils produced by different plant genera are inmany cases biologically active, endowed with antimicrobic, allelopathic, antioxidant andbio-regulatory properties. The antimicrobial abilities of essential oils, among which citrus oils are also shown to be a particularly interesting field for applications within the foodand cosmetic industries (Caccioni et al., 1998). Some essential oils were used in skincare products and for acne control. It is known that oil ofbergamot is receiving renewed popularity in aromatherapy. The peel of Citrus fruits is a rich source of flavanones and many polymethoxylatedflavones, which are very rare in other plants (Ahmad et al., 2006). These compounds, notonly play an important physiological and ecological role, but are also of commercialinterest because of their multitude of applications in the food and pharmaceuticalindustries. Naringin and hesperidin have many biological activities such as antioxidant, anti mutagenic effect, analgesic, anti-inflammatory. Citrus peel extracts showed a significant antibacterial activity against all the test organisms. The *Citrus lemon* and *Citrus sinensis* peels could act as anti-inflammatory, anticlotting, antioxidant, immune enhancers and hormone modulators. *Citrus lemon* and *Citrus sinensis* peels have high quantity of saponin which has haemolytic activity and cholesterol binding properties. Therefore, in addition to their use as drugs, citrus peels can be used as a food preservative or even as food Supplement as many literature says that they are highly nutritive (Giuseppe et al., 2007).

2.1 Uses and importance

Citrus aurantifolia/lime: -.The *Citrus aurantifolia* have antioxidant effect. The juice and peel extract of *Citrus aurantifolia* have antioxidant properties. Antioxidant effects of *Citrus aurantifolia* juice and peel extract on LDL (Low density lipoprotein) oxidation. The principal use is still for food, refreshing drinks, tasty disserts, and for seasoning meats, vegetables, salads, sauces, and casseroles (Ehler 2002, Katzer 2002). Essential oils of key lime and some other citrus fruits cause phytophotodermatosis in sensitive individuals (Bruneton 1999). Key lime is used to treat a huge number of ailments (Burkill 1997, Liogier, 1990). It is a good honey plant. The plant can be used for a living fence post (Little and Wadsworth, 1964) and can be formed into a hedge (Burkill, 1997).

Citrus maxima or *Citrus grandis*/pommelo: -.They used for hyperlipidemia, atherosclerosis, reducing hematocrit counts, cancer, psoriasis, and for weight loss and obesity .Seed extract is used orally for bacterial, viral, and fungal infections including yeast infections, *Giardia lamblia* and Entamoeba histolytica (Orwa et al., 2009). Oil is used for muscle fatigue, hair growth, toning the skin, and for acne and oily skin. It is also used for the common cold, swine flu, and flu (influenza). Seed extract is used topically as a facial cleanser, first-aid treatment, as a treatment for mild skin irritations, and as a vaginal douche for vaginal candidiasis (yeast infection). It is also used as an ear or nasal rinse for preventing and treating infections; as a gargle for sore throats; and a dental rinse for preventing gingivitis, promoting healthy gums, and as a breath freshener. In food and beverages, it is consumed as a fruit, juice, and is used as a flavoring component (Orwa et al., 2009).

Citrus limon/lemon: - It helps in production of white blood cells (WBC) and antibodies in blood which attacks the invading microorganism and prevents infection. Lemon is an antioxidant which deactivates the free radicals preventing many dangerous diseases like stroke, cardiovascular diseases and cancers. Lemon lowers blood pressure and an increase the levels of HDL (High density lipoprotein; good Cholesterol). It is found to be anti-carcinogenic which lowers the rates of colon, prostate and breast cancer (Gulsen and Roose, 2004). They prevent faulty metabolism in the cell, which can predispose a cell to becoming carcinogenic. Lemon juice is used to prevent common cold. Lemon juice is given to prevent/treat urinary tract infection Gonorrhea. Lemon juice relieves colic pain and gastric problems. Lemon juice soothes the dry skin when applied with little glycerin. Lemon juice used for marinating seafood or meat kills bacteria and other organisms present in them, thereby prevents many gastric-intestinal tract infections (Pnenniston et al., 2008).

3. Hypothesis

The current research work is based on the following hypothesis

1)   Antimicrobial properties of various honey shows difference in their action aganist pathogens.

2)   Lime juice also posses antimicrobial activity.

4. Materials and Methods

4.1 Study area

Kerala state covers an area of 38,863 km^2 with a population density of 859 per km^2 and spread across 14 districts. The climate is characterized by tropical wet and dry with average annual rainfall amounts to 2,817 ± 406 mm and mean annual temperature is 26.8°C (averages from 1871-2005; Krishnakumar et al., 2009).

Maximum rainfall occurs from June to September mainly due to South West Monsoon and temperatures are highest in May and November (Figure 1).

4.2 Sample collection

Sampling locations were selected in Kerala based on an elaborative baseline survey conducted during January 2015 to March 2015. The samples were collected based on an elaborative iterative survey as well as traditional knowledge from local people. Various honey samples were collected from different parts of Kerala, locations of the sample collection areas were recorded using a Trimble Geoexplorer II (Trimble Navigation Ltd, Sunnyvale, California) and data were transferred using GPS pathfinder Office software (Trimble Navigation Ltd, Sunnyvale, California).

Varieties of citrus are grow in south India. Selection of the plant samples was primarily based on its universal and local abundance. Three types of honey were collected; 'Cheruthen'-produced by bees belongs to the *Melipona irridipennis* species; Vanthen'- produced by bees belongs to the *Apis indica* species; 'Kattutthen'- produced by bees belongs to the *Apis dorsata* species.

4.3 Description of the species

Citrus aurantifolia leaves are green in colour. Leaves are 10 cm long. *Citrus maxima* leaves are dark green in colour. They appear simple and have 12 cm long leaves which are leathery. Adult leaves are used for DNA extraction of *Citrus maxima. Citrus limon* leaves are light green in colour. Leaves are in oval shape. They are about 8 cm in long.

4.3.1 Citrus aurantifolia / lime:

Lime is a shrubby tree, to 5 m, with many thorns. The trunk rarely grows straight, with many branches that often originate quite far down on the trunk. The leaves are ovate 1–3.5 in long, resembling orange leaves. The flowers are 1 in diameter, are yellowish white with a light purple tinge on the margins. Flowers and fruit appear throughout the year but are most abundant from May to September. Lime has an odour similar to lemon, but more fresh. The juice is as sour as lemon juice, but more aromatic. The English name lime originated from Arabia limun and Persian limou.

4.4 Microorganisms used in the study

4.4.1 Bacillus

Bacillus is a genus of gram positive, rod shaped (bacillus), bacteria and a member of the phylum Firmicutes. Bacillus species can be obligate aerobes (oxygen reliant), or facultative anaerobes (having the ability to be aerobic or anaerobic). They will test positive for the enzyme catalyse when there has been oxygen used or present (Turnbull et al., 1996). This gram positive bacterium is a common mesophilic, endospore forming saprophyte. This bacterium being thermoduric may cause concerns in food and dairy industries. Bacillus species may survive milk pasteurization or inadequate heat treatment during canning of foods. *Bacillus subtilis* has proved a valuable model for research. Other species of Bacillus are important pathogens, causing anthrax and food poisoning. This bacterium has been listed among food borne pathogens involved inoutbreaks from contaminated food and water. It produces rope in the open texture of bread, and is capable of spoilage in under processed foods (Madigen et al., 2005). It is known to produce an antibiotic bacitracin, which is used for topical treatment of infections caused by gram positive bacteria.

4.4.2 Klebsiella

Klebsiella is a genus of non motile, Gram-negative, oxidase negative, rod shaped bacteriawith a prominent polysaccharide based capsule (Ryanet al., 2004). It is named after the German microbiologist Edwin Klebs. Klebsiella species are ubiquitous in nature. This is thought to be due to distinct sub lineages developing specific niche adaptations, with associated biochemical adaptations which make them better suited to a particular environment. They can be found in water, soil, plants, insects, animals and humans (Ryan et al., 2004). Klebsiella bacteria tend to be rounder and thicker than other members of the Enterobacteriaceae family. They typically occur as straight rods with rounded or slightly pointed ends. They can be found singly, in pairs or in short chains. Diplobacillary forms are commonly found in vivo. They have no specific growth requirements and will grow well on standard laboratory medium, but grow best between 35 and 37°C and at pH 7.2. The species are facultative anaerobes, and most strains can survive with citrate and glucose as their sole carbon sources and ammonia as their sole nitrogen source (Ristuccia et al., 1984).

Members of the genus produce a prominent capsule, or slime layer, which can be used for serologic identification, but molecular serotyping may replace this method. Klebsiella species are routinely found in the human nose, mouth, and gastrointestinal tract as normal flora; however, they can also behave as opportunistic human pathogens (Ristuccia et al., 1984). Klebsiella species are known to also infect a variety of other animals, both as normal flora and opportunistic pathogens.

4.4.3 Micrococcus

Micrococcus is a genus of bacteria in the Micrococcus family. Micrococcus occurs in a wide range of environments, including water, dust, and soil. Micrococci have Gram positive spherical cells ranging from about 0.5 to 3 μm in diameter and typically appear in tetrads. They are catalase positive; oxidize positive, indole negative and citrate negative. Micrococcus has a substantial cell wall, which may comprise as much as 50% of the cell mass. The genome of Micrococcus is rich in guanine and cytosine (GC), typically exhibiting 65 to 75% GC content. Micrococci often carry plasmid (ranging from 1 to 100 MDa in size) that providesthe organism with useful traits.

Micrococci have been isolated from human skin, animal and dairy products, and beer. They are found in many other places in the environment, including water, dust, and soil. *Micrococcus luteus* on human skin transforms compounds in sweat into compounds with an unpleasant odour. Micrococci can grow well in environments with little water or high salt concentrations. Most are mesophiles; some, like *Micrococcus antarcticus* (found in Antarctica) are psychrophiles.

Though not a spore former, Micrococcus cells can survive for an extended period of time: unprotected cultures of soil micrococci have been revived after storage in a refrigerator for 10 years. *Micrococcus luteus* has survived for at least 34,000 to 170,000 years on the basis of 16S rRNA analysis, and possibly much longer (Agnihotri et al., 2004).

4.4.4 Staphylococcus

Staphylococcus is a gram positive, spherical bacteria occurring in short chains or grape like clusters. It is commonly found on the skin and in the nose of healthy people (Ryan et al., 2004). It produces16 numerous toxins that cause toxic-shock syndrome and staphylococcal scarlet fever. It is responsible for a number of diseases in humans like food poisoning, wound infections, skin infections, pneumonia, and toxic shock syndrome. It has been reported by the World Health Organization (WHO) that all over the world Staphylococcushas developed resistance to most potent antibiotics which were used against *Staphylococcus aureus* infections. 95% strains of *Staphylococcus aureus* are resistant to penicillin and 60% are resistant to methicillin. Apart from these antibiotics, it has also developed resistance to quinolone and vancomycin. The factors behind this resistant potential are numerous like resistances to penicillin are mediated by blaZ, the gene that encodes β lactamase. The

Staphylococcus genus includes at least 40 species. Of these, nine have two subspecies and one has three subspecies. Found worldwide, they are a small component of soil microbial flora (Madigan et al., 2005).

4.4.5 Escherichia coli

E. coli is a member of the large bacterial family, Enterobacteriaceae, the enteric bacteria, which are facultatively anaerobic, gram negative rods that live in the intestinal tracts of animals in health and disease. Pathogenic strains of *E. coli* are responsible for three types of infections in humans: urinary tract infections (UTI), neonatal meningitis, and intestinal diseases (gastroenteritis). Escherichia that is commonly found in the lower intestine of warm blooded organisms (endotherms). The strains can cause diarrhoea: enterotoxigenic, enteroinvasive, enteropathogenic, enterohaemorrhagic and enteroaggregative.

Some strains are emerging as potentially important causes of persistent diarrhoea in patients with acquired immune defficency syndrome (AIDS) and in children in tropical areas. Over 700 antigenic types (serotypes) of *E. coli* are recognized based on O, H, and K antigens. In northern India, 302 *E. coli* isolates from human and animal populations were checked for their antibiotic susceptibility and the results obtained showed the prevalence of multidrug resistant strains which accounted 41%. Other isolates were reported to be resistant to ampicillin (43.5%), oxytctracycline (36·4%) and trimethoprim sulphamethoxazole (9.3%). Similarly, in London, *E. coli* isolates from urine samples of different people were tested for their ability to resist various antibiotics, commonly used as empirical oral treatments for urinary tract infections. It was found that out of 11,865 isolates of *E. coli*, only 55% and 40% of isolates showed resistance to ampicillin and trimethoprim respectively. While 94% isolates were susceptible to nitrofurantoin followed by gentamicin (93.7%), and cefpodoxime (92%) (Dabbah et al., 1970). According to the US Centre for Disease Control and Prevention (CDC), *E. coli* is one of the leading causes of food borne illness in the US. Yearly, an estimated 76 million Americans fall ill from some form of food-borne illness, 325,000 lands in the hospital, and 5,000 die.

Most *E. coli* strains are harmless, but some serotypes can cause serious food poisoning in their hosts, and are occasionally responsible for product recalls due to contamination. The harmless strains are part of the normal flora of the gut, and can benefit their hosts by producing vitamin K2, and preventing colonization of the intestine with pathogenic bacteria (Zaika, 1975).

4.6 Preparation of samples

The honey sample used for the present studies was collected from honey producers. The honey samples 'Cheruthen,' 'Vanthen' were collected from Neeloor village, Kottayam District and 'Kattuthen' was collected from Poomala village, Idukki District. Lemon was collected from local market of Ramapuram, Kottayam District. The lime juice was extracted by squeezing and the concentrated extract was kept aseptically till usage.

4.7 Preparation of test organisms

The bacterial strains used for the test were Bacillus species, *Staphylococcus aureus*, *Escherichia coli*, Klebsiella species and Micrococcus species. The cultures have procured from the culture collection in the Biotechnology division of Mar Augusthinose College, Ramapuram Kottayam District. The test organisms were collected from selected strain of bacterium strain was maintained in nutrient agar slants.

4.8 Preparation of inoculam and culture media

4.8.1 Inoculam prepartion

Muller hinton agar (MHA) or nutrient broth was used as the media for the culturing of bacterial strains. Loopful of bacterial broth culture of each bacteria was inoculated inseparate nutrient broth and incubated at 37°C for 24 hours.

4.8.2 Preparation of nutrient agar slants

Prepared nutrient agar medium using the constituents, poured 8 ml of medium into test tube and plugged with cotton. Transferred all the tubes into a test tube stand and autoclaved at 121°C, at 15 lbs for 20 minutes. Sterilized tubes were taken out and placed them in slanting position by giving a support for agar slant. Then allowed the medium to solidify. Transfered all tubes to laminar airflow chamber. Marked each tube by name of microorganism to be inoculated. Aseptically inoculated each microbe on separate slant by using an inoculating loop. Incubated the test tubes at 37°C for 24 hours. Stored these tubes in refrigerator for further use.

4.8.3 Preparation of Muller hinton agar

Accurately weighted the chemical ingredients of the MHA and transferred into beaker containing 500 ml of distilled water. Gently heated the constituents with slight agitation, added distilled water to make the volume 1 litre. The pH was adjusted to 7.3 and poured the preparation to a conical flask.The mouth of the flaskswere plugged with cotton and covered with aluminium foil. Autoclaved at 121°C temperature, 15 lbs pressure for 15 minutes.

4.9 Antibacterial assay

Kirby-Bauer method was followed for disc diffusion assay. In vitro antimicrobial activity was screened by using MHA obtained from Hi Media (Mumbai). The MHA plates were prepared by pouring 25 ml of molten media into sterile petriplates. The plates were allowed to solidify for 20 minute and inoculum of each bacterial culturewas swabbed uniformly and the inoculums were allowed to dry for 5 min. The wells of 6 mm diameter were made on the plates using sterile tips. The different concentrations of samples were loaded on wells in the petriplates. The samples was allowed to diffuse for 5 minutes and the plates were kept for incubation at 37°C for 24 h. Microbial growth was determined by measuring the diameter of the zone of inhibition around each well. At the end of incubation, inhibition zones formed around the sample wells were measured with transparent ruler in millimetre. These studies were performed in triplicate.

4.10 Phytochemical analysis

Freshly prepared extracts were subjected to standard phytochemical analyses to find the presence of the following phytochemical constituents like phenols, flavanoids, alkaloids, glycosides, tannins, saponins and steroids.

4.10.1 Detection of alkaloid-Hager's test

Extract were dissolved separately in hydrochloric acid and filtered. Filtrate was treated with Hager's reagent (satured picric acid solution).

4.10.2 Detection of carbohydrates-Benedict's test

Extract were treated with 2 drops of alcoholic alpha naphthol solution in a test tube. Formation of the violet ring at the junction indicates the presence of carbohydrates. Filtrates were treated with 1ml of Benedict's reagent and heated gently.

4.10.3 Detection of glycosides-modified borntrager's test

Extract were hydrolysed with 2 drops of dilute hydrochloric acid (HCl), and then subjected to test for glycosides. Extract were treated with 2 drops of Ferric chloride ($FeCl_3$) solution and immersed in boiling water for about 5 minutes. The mixture was cooled and extracted with equal volume of benzene solution. The benzene layer was separated and treated with ammonia solution.

4.10.4 Detection of saponins-Foam test

0.5 g of extract was shaken with 2 ml of water.

4.10.5 Detection of phytosterols-Salkowski's test

Extracts were treated with 1 drops of chloroform and filtered. The filtrates were treated with few drops of concentrated sulphuric acid, shaken and allowed to stand.

4.10.6 Detection of phenols-Ferric chloride test

Extract were treated with 3 to 4 drops of $FeCl_3$ solution.

4.10.7 Detection of tannins-Gelatine test

To the extract, 1% gelatine solution containing sodium chloride (NaCl) was added.

4.10.8 Detection of flavanoids-Lead acetate test

Extract were treated with few drops of lead acetate solution.

4.10.9 Detection of protein and amminoacid-Xanthoproteic test

The extracts were treated with few drops of concentrated nitric acid (HNO_3).

4.10.10 Detection of diterpense-Copper acetate test

Extract were dissolved in water and treated with 3-4 drops of copper acetate ($C_2H_3CuO_2$) solution.

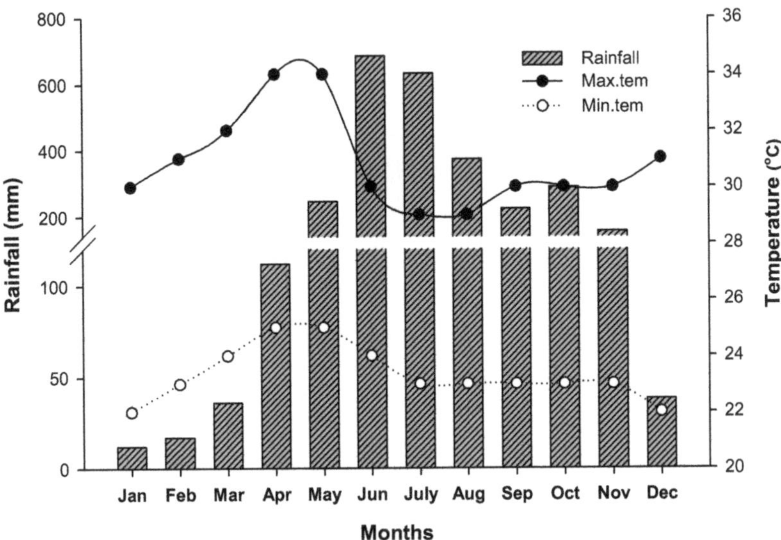

Figure 1. Mean monthly rainfall (mm), maximum and minimum temperatures (°C) in Kerala, India (1871-2005; Krishnakumar et al., 2009).

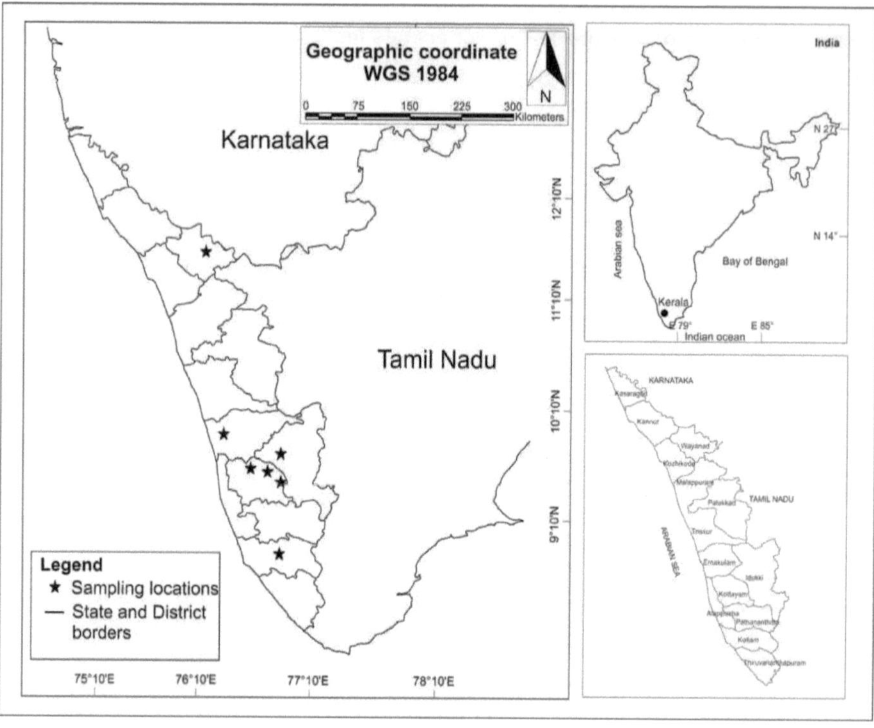

Figure 2. Map of Kerala showing the various sample collection points.

4.11 Statistical analysis

The survey results were analyzed and descriptive statistics were done using SPSS 12.0 (SPSS Inc., an IBM Company, Chicago, USA) and graphs were generated using Sigma Plot 7 (Systat Software Inc., Chicago, USA).

Figure 3.) Description of the various citrus varieties a), e) *Citrus aurantiifoli* tree bearing fruits, b) *Citrus maxima*, c) *Citrus limon*, d) orange fruit and blossom, f) pompia. Photo courtesy: Wikipedia.

Figure 4. Description of the various citrus varieties a) Lime fruits, b) amber sweet oranges, c) Clementine peeled, d) cross section of the various types of citrus fruits, e) citrus canker on fruit, f) odichukuthi lime cross section. Photo courtesy: Wikipedia.

.

Figure 5. Description of the various citrus varieties a) citrus fruit cut open, b) *Citrus australasica* fruit, c) citrus leaf, d) unripe mandarin fruit, e) odichukuthi lime. Photo courtesy: Wikipedia.

Figure 6. Key lime (*Citrus aurantifolia*) description a) ripe key lime fruit, b) mature key lime fruit, c) and d) key lime flowers. Photo courtesy: Wikipedia.

Figure 7. Pomelo (*Citrus maxima/Citrus grandis*) description a) mature *Citrus maxima* fruit, b) pomelo fruit cut open, c) *Citrus maxima* flower, d) *Citrus maxima* leaf, e) and f) pomelo flesh in red and cream colour. Photo courtesy: Wikipedia.

Figure 8. Leamon description a) leamon fruit and flower, b) leamon fruit on tree, c) leamon flower, d) leamon fruit cut open, e) leamon fruits on tree. Photo courtesy: Wikipedia.

Figure 9. Honey bee description a), b) and e) *Apis indica* collecting nectar and pollen from flowers, c) *Apis indica* worker emergence stages, d) *Apis indica* worker with pollen collected on pollen baskets, f) *Apis dorsata*. Photo courtesy: Wikipedia.

Figure 10. Honey bee description a) *Apis indica* brood, b) comb showing a day old egg, c) stages of development of egg and larvae, d) *Apis indica* brood cut open showing pupal stages, e) *Apis indica* worker foraging on flower. Photo courtesy: Wikipedia.

Figure 11. Honey bee description a) and b) *Apis dorsata* collecting nectar and pollen from flowers, c) *Apis dorsata* colony on tree, d) and e) stingless bee honey, f) stingless bee brood. Photo courtesy: Wikipedia (a, b, and c).

Figure 12. Honey bee description a) stingless bee colony entrance tube, b) stingless bee colony on woddenlog, c) stingless bee colony on wooden box, d) e) and f) stingless bees collecting nectar and pollen from various flowers.

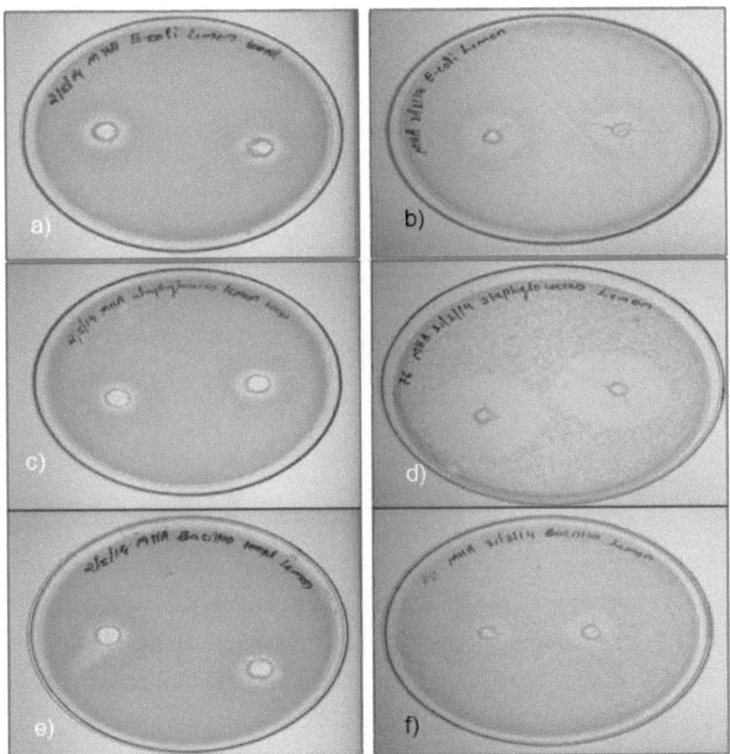

Figure 13. Zone of inhibition of lime juice extract aganist various bacterial isolates at different concentrations a) zone of inhibition prodcuced aganist *E.coli* using 100 µl extract, b) zone of inhibition prodcuced aganist *E.coli* using 200 µl extract, c) zone of inhibition prodcuced aganist *Staphylococus* using 100 µl extract, d) zone of inhibition prodcuced aganist *Staphylococus* using 200 µl extract, e) zone of inhibition prodcuced aganist *Bacillus* using 100 µl extract, f) zone of inhibition prodcuced aganist *Bacillus* using 200 µl extract.

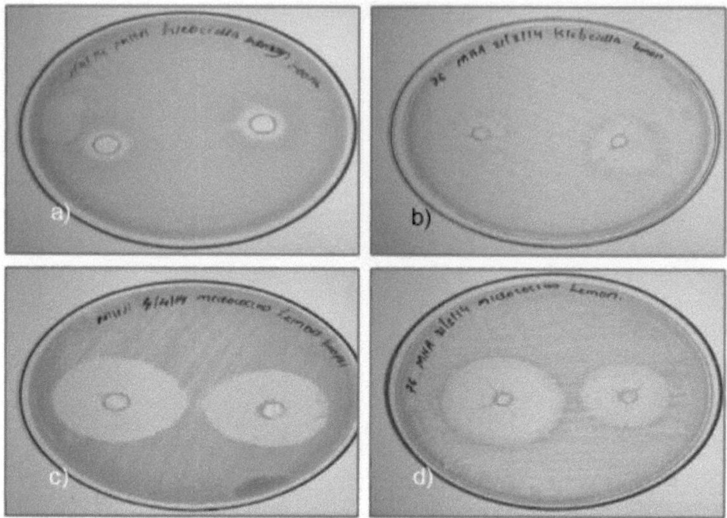

Figure 14. Zone of inhibition of lime juice extract aganist various bacterial isolates at different concentrations a) zone of inhibition prodcuced aganist *Klebsiella* using 100 µl extract, b) zone of inhibition prodcuced aganist *Klebsiella* using 200 µl extract, c) zone of inhibition prodcuced aganist *Micrococus* using 100 µl honey extract, d) zone of inhibition prodcuced aganist *Micrococus* using 200 µl extract.

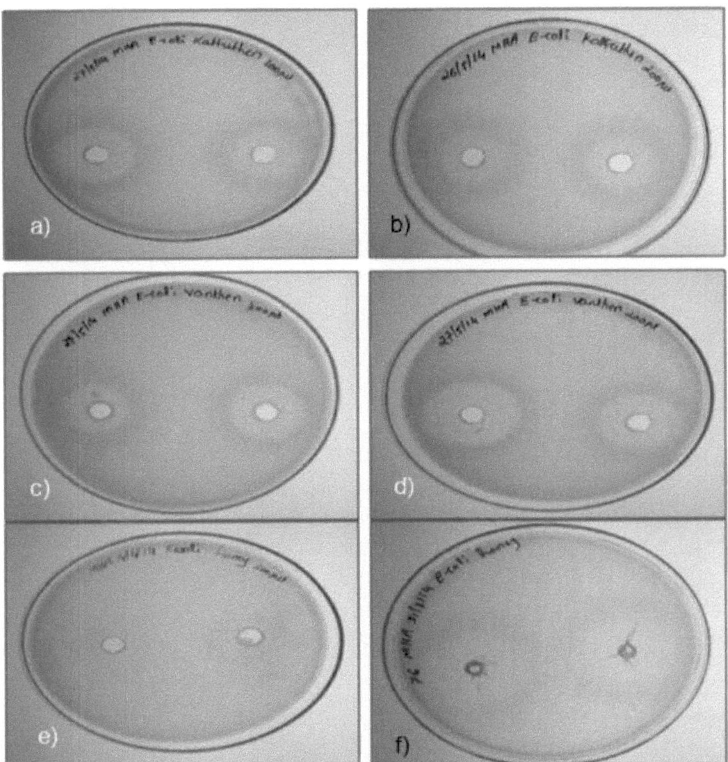

Figure 15. Zone of inhibition of various types of honey aganist *E.coli* at different concentrations a) zone of inhibition prodcuced aganist using 100 µl of kattuthen extract, b) zone of inhibition prodcuced aganist using 200 µl extract, c) zone of inhibition prodcuced aganist using 100 µl vantheen extract, d) zone of inhibition prodcuced aganist using 200 µl vantheen extract, e) zone of inhibition prodcuced aganist using 100 µl of cheruthen extract, b) zone of inhibition prodcuced aganist using 200 µl extract,.

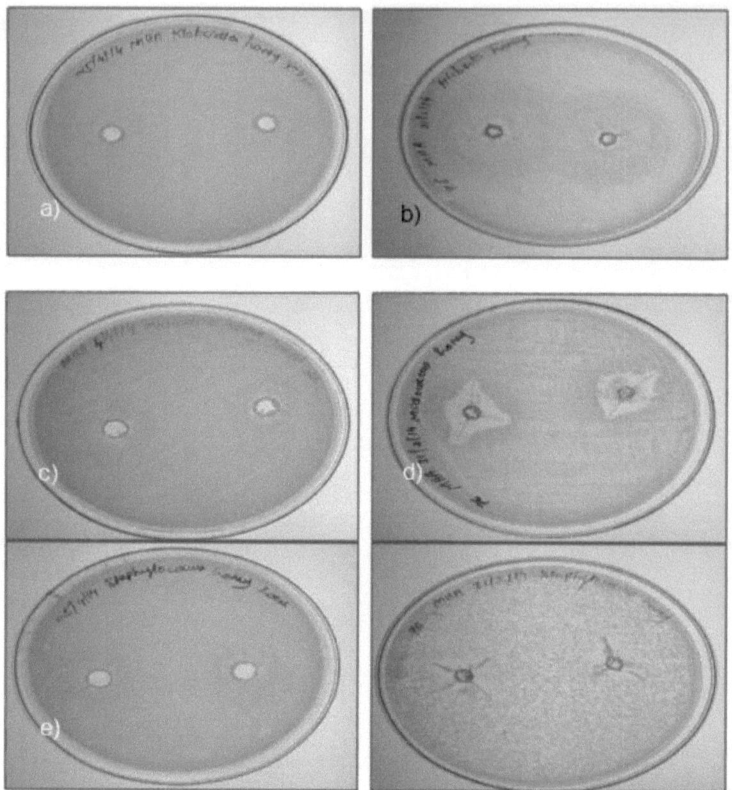

Figure 16. Zone of inhibition of cheruthen aganist various bacterial isolates at different concentrations a) zone of inhibition prodcuced aganist *Klebsiella* using 100 μl of honey extract, b) zone of inhibition prodcuced aganist *Klebsiella* using 200 μl honey extract, c) zone of inhibition prodcuced aganist *Micrococus* using 100 μl honey extract, d) zone of inhibition prodcuced aganist *Micrococus* using 200 μl honey extract, e) zone of inhibition prodcuced aganist *Staphylococus* using 100 μl of honey extract, b) zone of inhibition prodcuced aganist *Staphylococus* using 200 μl honey extract.

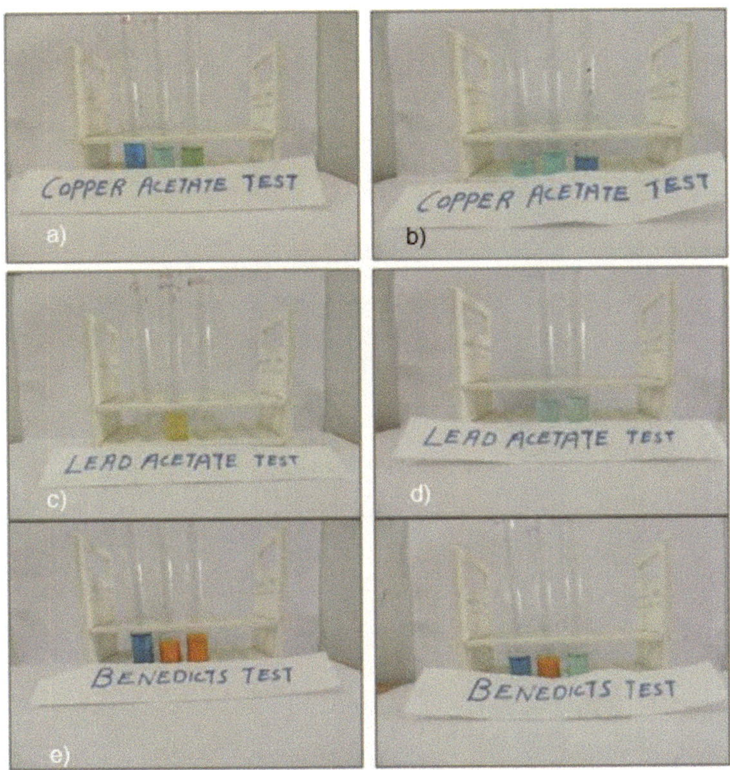

Figure 17. Phytochemical analysis of honey (vanthen and cheruthen) a) copper acetate test; tube 1: control, tube 2: vanthen, tube 3: cheruthen, b) copper acetate test; tube 1: kattuthen, tube 2: lime juice, tube 3: control, c) lead acetate test; tube 1: control, tube 2: cheruthen, tube 3: vanthen, d) lead acetate test; tube 1: lime juice, tube 2: kattuthen. e) Benedict's test; tube 1: control, tube 2: cheruthen, tube 3: vanthen, f) Benedict's test; tube 1: control, tube 2: kattuthen, tube 3: lime juice.

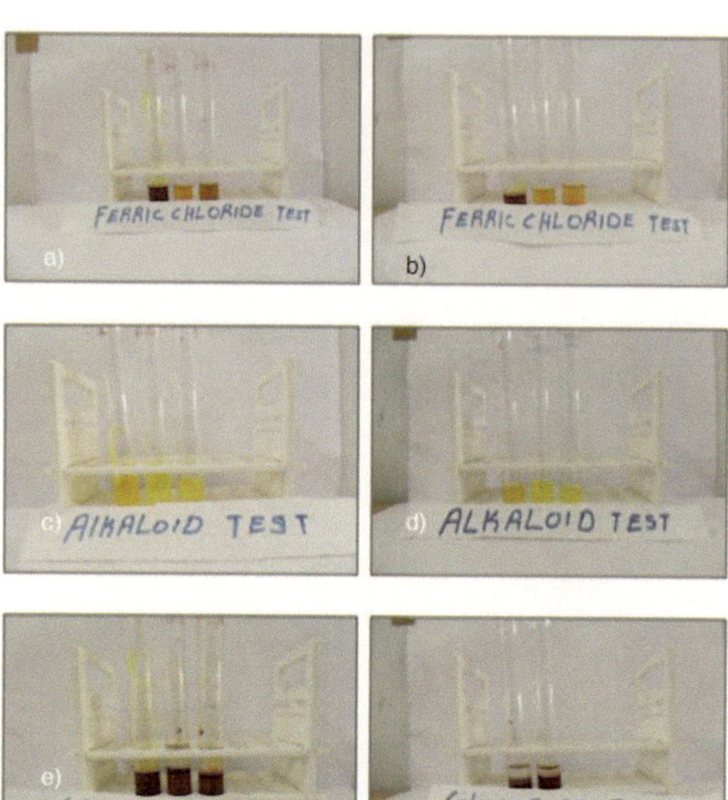

Figure 18. Phytochemical analysis of honey (vanthen and cheruthen) a) ferric chloride test; tube 1: control, tube 2: vanthen, tube 3: cheruthen, b) ferric chloride test; tube 1: control, tube 2: lime juice, tube 3: kattuthen, c) alkaloid test; tube 1: control, tube 2: vanthen, tube 3: cheruthen, d) alkaloid test; tube 1: control, tube 2: kattuthen, tube 3: lime juice, e) glycosides test; tube 1: control, tube 2: cheruthen, tube 3: vanthen, f) glycosides test; tube 1: lime juice, tube 2: kattuthen.

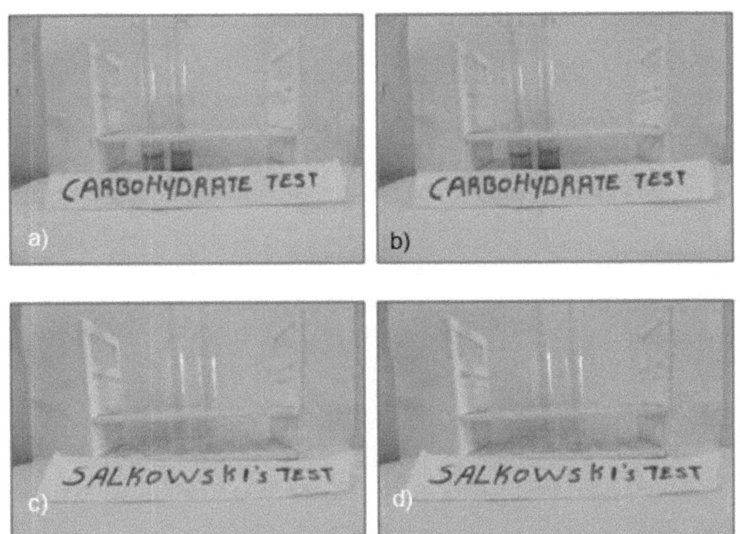

Figure 19. Phytochemical analysis of honey (vanthen and cheruthen) a) carbohydrate test; tube 1: cheruthen, tube 2: vanthen, b) carbohydrate test; tube 1: tube 1: lime juice, tube 2: kattuthen, c) Salkowski's test; tube 1: cheruthen, tube 2: vanthen, d) Salkowski's test; tube 1: kattuthen, tube 2: lime juice.

Table 1. Phytochemical analysis of various honey samples.

Sl.No.	Tests	Samples used			
		Lime juice	Cheruthen	Vanthen	Kattuthen
1.	Alkaloid (Hager's test)	+	-	+	+
2.	Carbohydrates (Benedict's test)	−	+	+	+
3.	Saponins (Foam test)	−	+	+	−
4.	Phytosterol's (Salkowski's test)	−	−	−	−
5.	Phenol's (Ferric chlorid test)	−	−	−	−
6.	Flavanoids (Lead acetate test)	+	+	+	+
7.	Tannins (Gelatine test)	+	+	+	+
8.	Protein and aminoacid (Xanthoproteic test)	−	−	−	−
9.	Diterpenes (Copper acetate test)	−	+	−	+
10.	Glycoside (Borntrager's test)	+	+	+	+

+ indicates presence; - indicates absence.

Table 2. Different vernacular names of *Citrus aurantifolia* (Key lime) in India.

Language	Names
Scientific names	*Citrus aurantifolia*
Name in various Indian languages	
Sanskrit	Matulunga, Nimbukah
Hindi	Kaghzi-nimbu, Nimbu
Bengali	Lebu
Marathi	Ambatanimbu, Limbu, Mavalanga
Kannada	
Konkani	
Malayalam	Naranga
Tamil	Champalam, Champiram
Telugu	Nimma

Table 3. Different vernacular names of *Citrus maxima* (Pomelo) in India.

Language	Names
Scientific names	*Citrus maxima*
	Citrus grandis
Name in various Indian languages	
Sanskrit	Karuna, Mallikapuspa
Hindi	Batawi-nimbu, Cakotara, Papar-mas
Bengali	Batabi lebu
Marathi	Bampara, Cakotara, Papannasa
Kannada	Cakota hannu
Konkani	Toranda
Malayalam	Kampilinarannga
Tamil	Metukku, Pampalimacu, Pommacu
Telugu	Pamparapanasa

Table 4. Different vernacular names of *Citrus limon* (Lemon) in India.

Language	Names
Scientific names	*Citrus limon*
Name in various Indian languages	
Sanskrit	Karuna, Mallikapuspa
Hindi	Nimbu
Bengali	
Marathi	
Kannada	
Konkani	
Malayalam	
Tamil	Elumicchai
Telugu	

Table 5. Zone of inhibition (mm) of lime juice and various honey samples against bacterial isolates (*E.coli*, Staphylococcus, Bacillus, Klebsiella and Micrococcus).

Bacterial isolates	Lime juice		Cheruthen		Vanthen		Kattuthen	
	Zone of inhibition (mm)		Zone of inhibition (mm)		Zone of inhibition (mm)		Zone of inhibition (mm)	
	100 µl	200 µl	100 µl	200 µl	100 µl	200 µl	100 µl	200 µl
E.coli	13.2 ± 0.83	16.6 ± 0.20	10.1 ± 0.13	30.2 ± 0.23	4.2 ± 0.53	8.2 ± 0.47	8.1 ± 0.61	12.5 ± 0.73
Staphylococcus species	13.5 ± 0.43	20.2 ± 0.28	13.4 ± 0.63	15.3 ± 0.23	3.5 ± 0.13	6.1 ± 0.13	7.5 ± 0.43	10.2 ± 0.44
Bacillus species	15.1 ± 0.21	15.3 ± 0.19	10.0 ± 0.41	25.2 ± 0.23	5.2 ± 0.40	7.2 ± 0.80	9.8 ± 0.77	13.4 ± 0.95
Klebsiella species	15.0 ± 0.28	16.5 ± 0.87	10.1 ± 0.73	30.1 ± 0.23	6.8 ± 0.43	8.9 ± 0.76	11.4 ± 0.48	14.1 ± 0.74
Micrococcus species	30.1 ± 0.63	32.1 ± 0.81	12.4 ± 0.20	17.3 ± 0.23	3.7 ± 0.23	6.7 ± 0.57	8.1 ± 0.23	10.5 ± 0.13

Numbers represent means ± one standard error (SE) of the mean

5. Results

5.1 Antimicrobial activity of honey and lime juice

Antimicrobial activities of lime juice and honey samples on various microorganisms were analysed. A sample volume 100 and 200 µl were used in the experiment and the zone of inhibition was measured.

5.1.1 Antimicrobial activity test using lime juice

Antimicrobial activity of lime juice against *E. coli* showed a zone of inhibition of 13.2 ± 0.83 mm, when 100 µl of lime juice is applied. When 200 µl of lime juice is applied the zone of inhibition was 16.6 ± 0.20 mm. Antimicrobial activity of lime juice against Staphylococcus species showed a zone of inhibition of 13.5 ± 0.43 mm, when 100 µl of lime juice is applied. When 200 µl of lime juice is applied the zone of inhibition was 20.2 ± 0.28 mm. Antimicrobial activity of lime juice against Bacillus species showed a zone of inhibition of 15.1 ± 0.21 mm, when 100 µl of lime juice is applied. When 200 µl of lime juice is applied the zone of inhibition was 15.3 ± 0.19 mm.

Antimicrobial activity of lime juice against Klebsiella species showed a zone of inhibition of 15.0 ± 0.28 mm, when 100 µl of lime juice is applied. When 200 µl of lime juice is applied the zone of inhibition was 16.5 ± 0.87 mm.

5.1.2 Antimicrobial activity test using kattuthen

Antimicrobial activity of Kattuthen against *E.coli* showed a zone of inhibition of 8.1 ± 0.61 mm, when 100 µl of kattuthen is applied. When 200 µl of kattuthen is applied the zone of inhibition was 12.5 ± 0.73 mm.

5.1.3 Antimicrobial activity test using vanthen

Antimicrobial activity of Vanthen against *E.coli* showed a zone of inhibition of 4.2 ± 0.53 mm, when 100 µl of vanthen is applied. When 200 µl of vanthen is applied the zone of inhibition was 8.2 ± 0.47 mm.

5.1.4 Antimicrobial activity test using cheruthen

Antimicrobial activity of Cheruthen against Klebsiella species showed a zone of inhibition of 10.1 ± 0.73 mm, when 100 µl of Cheruthen is applied. When 200 µl of Cheruthen is applied the zone of inhibition was 30.1 ± 0.23 mm. Antimicrobial activity of Cheruthen against *E.coli* showed a zone of inhibition of 10.1 ± 0.13 mm, when 100 µl of cheruthen is applied. When 200 µl of cheruthen is applied the zone of inhibition was 30.2 ± 0.23 mm.

Antimicrobial activity of Cheruthen against Micrococcus species showed a zone of inhibition of 12.4 ± 0.20 mm, when 100 µl of cheruthen is applied. When 200 µl of cheruthen is applied the zone of inhibition was 17.3 ± 0.23 mm.

Antimicrobial activity of Cheruthen against Staphylococcus species showed a zone of inhibition of 13.4 ± 0.63 mm, when 100 µl of cheruthen is applied. When 200 µl of cheruthen is applied the zone of inhibition was 15.3 ± 0.23 mm.

5.2 Phytochemical analysis

5.2.1 Detection of alkaloid-Hager's test

Presence of alkaloid confirmed by the formation of yellow colored precipitate.

5.2.2 Detection of carbohydrates-Benedict's test

Orange red precipitate indicated the presence of reducing sugars.

5.2.3 Detection of glycosides-modified borntrager's test

Formation of rose pink colour in the ammoniac layer indicated the presence of anthranol glycosides.

5.2.4 Detection of saponins-Foam test

Foam produced persists for 10 minutes and it indicated the presence of saponins.

5.2.5 Detection of phytosterols-Salkowski's test

Appearance of golden yellow colour indicated the presence of triterpenes.

5.2.6 Detection of phenols-Ferric chloride test

Formation of bluish black colour indicated the presence of phenols.

5.2.7 Detection of tannins-Gelatine test

Formation of white precipitate indicated the presence of tannins.

5.2.8 Detection of flavanoids-Lead acetate test

Formation of yellow colour precipitate indicated the presence of flavanoids.

5.2.9 Detection of protein and amminoacid-Xanthoproteic test

Formation of yellow colour precipitate indicated the presence of proteins.

5.2.10 Detection of diterpense-Copper acetate test

Formation of emerald green colour indicated the presences of diterpenes.

6. Discussion

The results showed that the lime juice possess antimicrobial activity against Klebsiella species, Micrococcus species, Staphylococcus species, Bacillus species, and *E coli.*

Lime juice had shown inhibition zone at 100 µl on Micrococcus species, is 30.1 ± 0.63 mm in diameter, *E coli* is 13.2 ± 0.83 mm in diameter, Klebsiella species is 15.0

± 0.28 mm in diameter, Bacillus species is 15.1 ± 0.21 mm in diameter and Staphylococcus species, and is 13.5 ± 0.43 mm in diameter. Lime juice had shown inhibition zone at 200 µl on micrococcus species is 32.1 ± 0.81 mm in diameter, *E coli* species is 16.6 ± 0.20 mm in diameter, Klebsiella species is 16.5 ± 0.87 mm in diameter, Bacillus species is 15.3 ± 0.19 mm in diameter and staphylococcus species is 20.2 ± 0.28 mm in diameter.

Cheruthen had shown inhibition zone at 100 µl on Micrococcus species is 12.4 ± 0.20 mm in diameter, *E coli* is 10.1 ± 0.13 mm in diameter, Klebsiella species is 10.1 ± 0.73 mm in diameter, Bacillus species is 10.0 ± 0.41 mm in diameter and Staphylococcus species and is 13.4 ± 0.63 mm in diameter. Cheruthen had shown inhibition zone at 200 µl on Micrococcus species is 17.3 ± 0.23 mm in diameter, *E coli* is 30.2 ± 0.23 mm in diameter, Klebsiella species is 30.1 ± 0.23 mm in diameter, Bacillus species is 25.2 ± 0.23 mm in diameter and Staphylococcus species is 15.3 ± 0.23 mm in diameter.

After this, the result obtained for lime juice, cheruthen, vanthen and kattuthen showed antimicrobial sensitivity property. Also by phytochemical screening, the phytochemicals present in lime juice, Cheruthen, vanthen, and kattuthen can be examined. Studies are in progress to further evaluate the mechanism of action of lime juice and honey sample and some organism associated with human diseases. We can use these compounds to cure disease causing microorganisms which are sensitive to phytochemical compounds in lime juice and honey.

The phytochemical analysis of lime juice contains the presence of alkaloids, flavonoids, tannins and glycosides. Cheruthen contains the phytochemicals like carbohydrates, flavonoids, tannins, saponins, glycosides and diterpenes. Vanthen contains the phytochemicalslike alkaloids, carbohydrates, flavonoids, saponins, tannins and glycosides. Andanalysis of Kattuthen revealed the presence of alkaloids, carbohydrates, flavonoids, tannins and glycosides.

7. Conclusions

This study was to analyse the samples viz. lime juice, Cheruthen, Vanthen, and Kattuthen for its antimicrobial property and also for the presence of different phytochemicals. The result showed that the samples have antimicrobial activity. Also the phytochemical examination of lime juice and honey samples showed that different types ofphytochemical substances are present in both lime juice and different types of honey samples. Further studies are required to reveal the role of each phytochemical and its contribution to the antimicrobial properties of the samples included in this study.

Acknowledgements

The authors are grateful for the cooperation of the management of Mar Augsthinose college for necessary support. Technical assistance from Binoy A Mulanthra is also acknowledged.

References

Abdul, W. A., Joseph, P. C., & Kennedy, H. E. (2012). Nutraceutical values of natural honey and its contribution to human health and wealth, *Nutrition and Metabolism, 9*(61), 1-12.

Adeleke, O. E, Olaitan, J. O., & Okepekpe, E. I., (2006). Comparative antibacterial activity of honey and gentamicin against *Escherichia coli* and *Pseudomonas aeruginosa. Annals Burn Fire Disasters, 19*(4), 201-204.

Agnihotri, V. K., Thappa, R. K., Meena, B., Kapahi, B. K., Saxena, R. K., Qazi, G. N., & Agarwal, S. G. (2004). Essential oil composition of aerial parts of *Angelica glauca* growing wild in North-West Himalaya (India).*Phytochemistry, 65*(16), 2411-2413.

Ahmad, M. M., Iqbal, Z., Anjum, F. M., & Sultan, J. I. (2006). Genetic variability to essential oil composition in four citrus fruit species. *Pakistan Journal of Botany, 38*(2), 319-324.

Akinmoladun, A. C., Ibukun, E. O., Afor, E., Obuotor, E. M., & Farombi, E. O. (2007). Phytochemical constituent and antioxidant activity of extract from the leaves of *Ocimum gratissimum. Scientific Research and Essays, 2*(5), 163-166.

Allen, K. L., Molan, P. C., & Reid, G. M. (1991). A survey of the antibacterial activity of some New Zealand honeys. *Journal of Pharmacy and Pharmacology, 43*(12), 817-822.

Al-Shuneigat, J., Cox, S. D., & Markham, J. L. (2005). Effects of a topical essential oil-containing formulation on biofilm-forming coagulase-negative staphylococci. *Letters in Applied Microbiology, 41*(1), 52-55.

Al-Waili, N., Akmal, M., Al-Waili, F., Saloom, K., & Ali, A. (2005). The antimicrobial potential of honey from United Arab Emirates on some microbial isolates. *Medical Science Monitor, 11*(12), 433-438.

Aronson, J. K. (2001). Forbidden fruit. *Nature medicine, 7*(1), 29-30.

Arshad Hussein, 2003. Phytochemical analysis and chemical composition of different branded and unbranded honey samples. *InternationalJournal of Microbiological Research, 4*(2), 132-137.

Barkley, N. A., Roose, M. L., Krueger, R. R., & Federici, C. T. (2006). Assessing genetic diversity and population structure in a citrus germplasm collection utilizing simple sequence repeat markers (SSRs). *Theoretical and Applied Genetics, 112*(8), 1519-1531.

Baur, A. W., Kirby, W. M., Sherries, J. C., Truck, M. (1966). Antibiotic susceptibility testing by a standard single disk method. *American Journal of Clinical Pathology, 45*(1), 493-496.

Bertoncelj, J., Doberšek, U., Jamnik, M., & Golob, T. (2007). Evaluation of the phenolic content, antioxidant activity and colour of Slovenian honey. *Food Chemistry, 105*(2), 822-828.

Blasa, M., Candiracci, M., Accorsi, A., Piacentini, M. P., Albertini, M. C., & Piatti, E. (2006). Raw Millefiori honey is packed full of antioxidants. *Food Chemistry, 97*(2), 217-222.

Bosi, G., & Battaglini, M. (1978). Gas chromatographic analysis of free and protein amino acids in some unifloral honeys. *Journal of Apicultural Research, 17*(3), 152-166.

Brown D, 1995. The herb society of America- New Encyclopedia of Herbs and their uses. *Darling, Kingsley Limited, London,* UK.

Bruneton, J. (1999). Toxic plants dangerous to humans and animals. Intercept Limited.

Bugno, A., Nicoletti, M. A., Almodóvar, A. A., Pereira, T. C., & Auricchio, M. T. (2007). Antimicrobial efficacy of *Curcuma zedoaria* extract as assessed by linear regression compared with commercial mouthrinses. *Brazilian Journal of Microbiology, 38*(3), 440-445.

Burkill, H. M. (1995). *The useful plants of west tropical Africa, Vols. 1-3* (No. 2. ed.). Royal Botanic Gardens, Kew.

Burt, S. A. (2004). Essential oils: their antibacterial properties and potential applications in foods: A review. *International Journal of Food Microbiology, 94*(3), 223-253.

Caccioni, D. R., Guizzardi, M., Biondi, D. M., Renda, A., & Ruberto, G. (1998). Relationship between volatile components of citrus fruit essential oils and antimicrobial action on *Penicillium digitatum* and *Penicillium italicum. International Journal of Food Microbiology, 43*(1), 73-79.

Chick, H., Shin, H. S., & Ustunol, Z. (2001). Growth and acid production by lactic acid bacteria and bifidobacteria grown in skim milk containing honey. *Journal of Food Science, 66*(3), 478-481.

Chirife, J., Scarmato, G., & Herszage, L. (1982). Scientific basis for use of granulated sugar in treatment of infected wounds. *The Lancet, 319*(8271), 560-561.

Cos, P., Vlietinck, A. J., Berghe, D. V., & Maes, L. (2006). Anti-infective potential of natural products: how to develop a stronger in vitro 'proof-of-concept'. *Journal of ethnopharmacology, 106*(3), 290-302.

Cvetnic, Z.., & Vladimir-Knezevic, S. (2004). Antimicrobial activity of grapefruit seed and pulp ethanolic extract. *Acta Pharm, 54*(3), 243-250.

da C Azeredo, L., Azeredo, M. A. A., De Souza, S. R., & Dutra, V. M. L. (2003). Protein contents and physicochemical properties in honey samples of *Apis mellifera* of different floral origins. *Food Chemistry, 80*(2), 249-254.

Dabbah, R., Edwards, V. M., & Moats, W. A. (1970). Antimicrobial action of some citrus fruit oils on selected food-borne bacteria. *Applied Microbiology, 19*(1), 27-31.

Davis, Leonard. *Basic methods in molecular biology*. Elsevier, 2012.

de Moraes Pultrini, A., Galindo, L. A., & Costa, M. (2006). Effects of the essential oil from *Citrus aurantium* L. in experimental anxiety models in mice. *Life Sciences, 78*(15), 1720-1725.

Dorman, H. J. D., & Deans, S. G. (2000). Antimicrobial agents from plants: antibacterial activity of plant volatile oils. *Journal of applied microbiology,88*(2), 308-316.

Duthie, G., & Crozier, A. (2000). Plant-derived phenolic antioxidants. *Current Opinion in Lipidology, 11*(1), 43-47.

Ebrahimzadeh, M. A., Hosseinimehr, S. J., & Gayekhloo, M. R. (2004). Measuring and comparison of vitamin C content in citrus fruits: introduction of native variety. *Chemistry: An Indian Journal, 1*(9), 650-652.

Ehler, J. T. (2002). Key lime (Citrus aurantifolia). Food Reference, Key West, FL.

Ekwenye, U. N., & Edeha, O. V. (2010). The antibacterial activity of crude leaf extract of *Citrus sinensis* (Sweet Orange). *International Journal of Pharmacy and Bio Sciences, 1*(4), 742-750.

FAO. (2008). FAOSTAT. Satistical database

Fernandez-Lopez, J., Zhi, N., Aleson-Carbonell, L., Perez-Alvarez, J. A., & Kuri, V. (2005). Antioxidant and antibacterial activities of natural extracts: application in beef meatballs. *Meat Science, 69*(3), 371-380.

Ferreres, F., García-Viguera, C., Tomás-Lorente, F., & Tomás-Barberán, F. A. (1993). Hesperetin: A marker of the floral origin of citrus honey. *Journal of the Science of Food and Agriculture, 61*(1), 121-123.

Friedman, M., Henika, P. R., & Mandrell, R. E. (2002). Bactericidal activities of plant essential oils and some of their isolated constituents against *Campylobacter jejuni, Escherichia coli, Listeria monocytogenes,* and *Salmonella enterica. Journal of Food Protection, 65*(10), 1545-1560.

Gattuso, G., Barreca, D., Gargiulli, C., Leuzzi, U., & Caristi, C. (2007). Flavonoid composition of citrus juices. *Molecules, 12*(8), 1641-1673.

Gharagozloo, M., Doroudchi, M., & Ghaderi, A. (2002). Effects of Citrus aurantifolia concentrated extract on the spontaneous proliferation of MDA-MB-453 and RPMI-8866 tumor cell lines. *Phytomedicine, 9*(5), 475-477.

Gheldof, N., Wang, X. H., & Engeseth, N. J. (2002). Identification and quantification of antioxidant components of honeys from various floral sources. *Journal of Agricultural and Food Chemistry, 50*(21), 5870-5877.

Gulsen, O., & Roose, M. L. (2001). Lemons: diversity and relationships with selected Citrus genotypes as measured with nuclear genome markers. *Journal of the American Society for Horticultural Science, 126*(3), 309-317.

Hammer, K. A., Carson, C. F., & Riley, T. V. (1999). Antimicrobial activity of essential oils and other plant extracts. *Journal of Applied Microbiology, 86*(6), 985-990.

Hayes, A. S., & Markovic, B. (2002). Toxicity of Beak *housie citrodora. (Lemon Myrthle).* Anti-microbial and *in vitro* cytotoxicity. *Food Chemistry and Toxicology, 40*(4), 535-543.

Hilu, K. W. L. A., & Liang, H. (1997). The matK gene: sequence variation and application in plant systematics. *American Journal of Botany, 84*(6), 830-830.

Ibrahim, Y.K., & Ogunmodede, M.S. (1991). Growth and survival of *Pseudomonas aeruginosain* some aromatic waters. *Pharmaceutical act a Helvetica, 66*(9-10), 286-288.

Imade, G. E., Sagay, A. S., Onwuliri, V. A., Egah, D. Z., Potts, M., & Short, R. V. (2005). Use of lemon or lime juice douches in women in Jos, Nigeria. *Sex Health, 2*(1), 237-239.

Jayaprakasha, G. K., & Patil, B. S. (2007). In vitro evaluation of the antioxidant activities in fruit extracts from citron and blood orange. *Food Chemistry, 101*(1), 410-418.

John K Francis

Johnson, L. A., & Soltis, D. E. (1994). matK DNA sequences and phylogenetic reconstruction in *Saxifragaceae s. str. Systematic Botany, 19*(1), 143-156.

Kanaze, F. I., Termentzi, A., Gabrieli, C., Niopas, I., Georgarakis, M., & Kokkalou, E. (2008). The phytochemical analysis and antioxidant activity assessment of orange peel (*Citrus sinensis*) cultivated in Greece-Crete indicates a new commercial source of hesperidin. *Biomedical Chromatography, 23*(1), 239-249.

Kandil, A., El-Banby, M., Abdel-Wahed, K., Abdel-Gawwad, M., & Fayez, M. (1987). Curative properties of true floral and false non floral honeys and induced gastric ulcers. *Journal of Drug Research of Egypt, 17*(1), 103-106.

Katzer, G. (2002). Lime [*Citrus aurantifolia*

Kawaii, S., Tomono, Y., Katase, E., Ogawa, K., Yano, M., Koizumi, M., & Furukawa, H. (2000). Quantitative study of flavonoids in leaves of Citrus plants. *Journal of Agricultural and Food Chemistry, 48*(9), 3865-3871.

Keles, O.S., Bakirel, A. T., & Alpinar, K. (2001). Screening of some Turkish plants for antibacterial activity. *Turkish Journal of Veterinary and Animal Sciences, 25*(4), 559-565.

Lindsey, K., Jäger, A. K., Raidoo, D. M., & van Staden, J. (1998). Screening of plants used by Southern African traditional healers in the treatment of dysmenorrhoea for prostaglandin-synthesis inhibitors and uterine relaxing activity. *Journal of Ethnopharmacology, 64*(1), 9-14.

Liogier, A. H., & Martorell, L. F. (2000). *Flora of Puerto Rico and adjacent islands: a systematic synopsis.* La Editorial, UPR.

Little Jr, E. L., & Frank, H. Wadsworth. 1964. Common trees of Puerto Rico and the Virgin Islands. *US Department of Agriculture, Agriculture Handbook, 249*, 428-429.

Lusby, P. E., Coombes, A. L., & Wilkinson, J. M. (2005). Bactericidal activity of different honeys against pathogenic bacteria. *Archives of Medical Research, 36*(5), 464-467.

Mabberley, D. J. (2008). *Mabberley's plant-book: a portable dictionary of plants, their classifications and uses* (No. Ed. 3). Cambridge University Press, London, UK.

Madigan M, Martinko J, ed. 2005. Brock Biology of Microorganisms(11th Ed.). *Prentice Hall. ISBN* 0-13-144329-1.

Mandal, M. D., & Mandal, S. (2011). Honey: its medicinal property and antibacterial activity. *Asian Pacific Journal of Tropical Biomedicine, 1*(2), 154-160.

Manthey, J. A., & Grohmann, K. (1996). Concentrations of hesperidin and other orange peel flavonoids in citrus processing byproducts. *Journal of Agricultural and Food Chemistry, 44*(3), 811-814.

Manthey, J. A., & Grohmann, K. (2001). Phenols in citrus peel byproducts. Concentrations of hydroxycinnamates and polymethoxylated flavones in citrus peel molasses. *Journal of Agricultural and Food Chemistry, 49*(7), 3268-3273.

Michaud, J. P. (1999). Aggregation by alatae of Toxoptera citricida (Homoptera: Aphididae). *Environmental entomology, 28*(2), 205-211.

Molan, P. C. (1992). The antibacterial activity of honey: 1. The nature of the antibacterial activity. *Bee world, 73*(1), 5-28.

Molan, P. C. (1992). The antibacterial activity of honey: 2. Variation in the potency of the antibacterial activity. *Bee world, 73*(2), 59-76.

Molan, P. C. (2001). Why honey is effective as a medicine. 2. The scientific explanation of its effects. In: Munn P, Jones R, editors. Honey and Healing. UK: *International Bee Research Association;* 59-76.

Moro. C., & Basile, G. (2000). Obesity and medicinal plants. *Fitoterapia.* (Supplement 1); 73-78.

Morton, J. F. (1987). *Fruits of warm climates.* JF Morton.

Nagi, T., Inoue, R., Kanamori, N., Suzuki, N., & Nagashima, T. (2006). Characterization of honey from different floral sources. Its functional properties and effects of honey species on storage of meat. *Food Chemistry, 97*(2), 256-262.

Natarajan, S. C., Williamson, D., Grey, J., Harding, K. G., & Cooper, R. A. (2001). Healing of an MRSA-colonized, hydroxyurea-induced leg ulcer with honey. *Journal of Dermatological Treatment, 12*(1), 33-36.

Nzeako, B. C., & Hamdi, J. (2000). Antimicrobial potential of honey on some microbial isolates. *Medical Sciences, 2*(2), 75-79.

Ortuño, A. M., Arcas, M. C., Benavente-Garcia, O., & Del Rio, J. A. (1999). Evolution of polymethoxy flavones during development of tangelo Nova fruits. *Food chemistry, 66*(2), 217-220.

Ortuño, A., Báidez, A., Gómez, P., Arcas, M. C., Porras, I., García-Lidón, A., & Del Rio, J. A. (2006). *Citrus paradisi* and *Citrus sinensis* flavonoids: Their influence in the defence mechanism against *Penicillium digitatum*. *Food Chemistry, 98*(2), 351-358.

Penjor, T., Yamamoto, M., Uehara, M., Ide, M., Matsumoto, N., Matsumoto, R., & Nagano, Y. (2013). Phylogenetic relationships of Citrus and its relatives based on matK gene sequences. *PloS one, 8*(4), e62574.

Prabuseenivasan, S., Jayakumar, M., & Ignacimuthu, S. (2006). In vitro antibacterial activity of some plant essential oils. *BMC Complementary and Alternative Medicine, 6*(1), 1-8.

Ryan, K. J., & Ray, C. G. (2004). *Sherri's Medical Microbiology* (4th Ed.). McGraw Hill. ISBN 0-8385-8529-9.

Seymour, F. I., & West, K. S. (1951). Honey--its role in medicine. *Medical times, 79*(2), 104-108.

Shahnah, S. M., Ali, S., Ansari, H., & Bagri, P. (2007). New sequiterpene derivative from fruit peel of *Citrus limon* (Linn) Burn. *F. Scientia Pharmaceutica, 75*(1), 165-170.

Siddiqi, M., Tricker, A. R., & Preussmann, R. (1988). Formation of N-nitroso compounds under simulated gastric conditions from Kashmir foodstuffs. *Cancer Letters, 39*(3), 259-265.

Singleton, P. (1999). *Bacteria in Biology, Biotechnology and Medicine* (5th Ed). Wiley.pp.444-454. ISBN 0-471-98880-4.

Sohn, H. Y., Son, K. H., Kwon, C. S., Kwon, G. S., & Kang, S. S. (2004). Antimicrobial and cytotoxic activity of 18 prenylated flavonoids isolated from medicinal plants: *Morus alba* L., *Morus mongolica* Schneider, *Broussnetia papyrifera* (L.) Vent, *Sophora flavescens* Ait and *Echinosophora koreensis* Nakai. *Phytomedicine, 11*(7), 666-672.

Sokovic, M., Marin, P. D., Brkic, D., & van Griensven, L. J. (2008). Chemical composition and antibacterial activity of essential oils against human pathogenic bacteria. *Food, 1*(2), 220-226.

Swingle, W. T., Reece, P. C., Reuther, W., Webber, H. J., & Batchelor, L. D. (1967). The citrus industry. *The citrus industry, 1.*

Tepe, B., Daferera, D., Sökmen, M., Polissiou, M., & Sökmen, A. (2004). In vitro antimicrobial and antioxidant activities of the essential oils and various extracts of *Thymus eigii. Journal of Agricultural and Food Chemistry, 52*(5), 1132-1137.

Wahdan, H. A. L. (1998). Causes of the antimicrobial activity of honey. *Infection, 26*(1), 26-31.

Weston, R. J. (2000). The contribution of catalase and other natural products to the antibacterial activity of honey: a review. *Food Chemistry, 71*(2), 235-239.

White, J. W. (1979). Composition of honey. In: Crane E, ed. *Honey: A comprehensive survey.* Heinemann, London: chap 5.

Wilson, C. L., Droby, G. G. (2000). Microbial Food Contamination, *CRC Press, Boca Raton, FL,* 1-304 (149-171).

Zaika, L. L. (1988). Spices and herbs: their antimicrobial activity and its determination. *Journal of Food Safety, 9*(2), 97-118.

Zaika, L.L, 1975. "Spices and herbs: their antimicrobial activity and its determination" *J. Food Safety,* 9:97-118.

Zumla, A., & Lulat, A. (1989). Honey--a remedy rediscovered. *Journal of the Royal Society of Medicine, 82*(7), 384-385.

YOUR KNOWLEDGE HAS VALUE

- We will publish your bachelor's and master's thesis, essays and papers

- Your own eBook and book - sold worldwide in all relevant shops

- Earn money with each sale

Upload your text at www.GRIN.com and publish for free